Drilling for Oil

Kate McArthur

NELSON
A Cengage Company

Australia • Brazil • Japan • Korea • Mexico • Singapore • Spain • United Kingdom • United States

Drilling for Oil

Text: Kate McArthur
Editor: Rebecca Crisp
Design: James Lowe
Series design: James Lowe
Photo researcher: Corrina Tauschke
Production controller: Adam Bextream
Reprint: Siew Han Ong

Acknowledgements
The author and publisher would like to acknowledge permission to reproduce material from the following sources:
Corbis/Amos Nachoum: p. 10; Corbis/Bob Fleumer/zefa: p. 21 (bottom); Corbis/Charles E. Rotkin: p. 3; Corbis/G. Bowater: p. 14; Corbis/Georgia Lowell: p. 21 (top); Corbis/Keith Wood: p. 15; Corbis/Peter Turnley: p. 9; Corbis/Roger Ressmeyer: p. 11; Corbis/Zhao Cheng An/Redlink: p. 6; Getty Images: pp. 16, 18–19, 22 (bottom right); HARALD SUND MMVIII/All rights reserved/SEATTLE, WASHINGTON USA: p. 20; iStockphoto: p. 22 (bottom left); Photolibrary/Ken Graham: pp. 1, cover; Photolibrary/Nicholas Secor: back cover; Photolibrary/Robert Harding: p. 7; Photolibrary/Science Photo Library/KARIM AGABI/EURELIOS: p. 12; Photolibrary/Science Photo Library/Paul Rapson: p. 4.

Every effort has been made to trace and acknowledge copyright. However, if any infringement has occurred, the publishers tender their apologies and invite the copyright holders to contact them.

Fast Forward Independent Texts
Level 22

For product information and technology assistance,
in Australia call 1300 790 853;
in New Zealand call 0508 635 766

For permission to use material from this text or product,
please email **aust.permissions@cengage.com**

ISBN 978 0 17 017992 8
ISBN 978 0 17 017899 0 (set)

Cengage Learning Australia
Level 7, 80 Dorcas Street
South Melbourne, Victoria Australia 3205

Cengage Learning New Zealand
Unit 4B Rosedale Office Park
331 Rosedale Road, Albany, North Shore NZ 0632

For learning solutions, visit **cengage.com.au**

Printed in China by 1010 Printing International Ltd
2 3 4 5 6 7 15

Drilling for Oil

Kate McArthur

Contents

CHAPTER 1

The Origins of Oil

Oil is a thick, dark-coloured liquid. It was formed over millions of years from the **remains** of animals and plants that lived in the sea.

When these plants and animals died, their remains dropped to the bottom of the ocean and were covered in layers of mud.

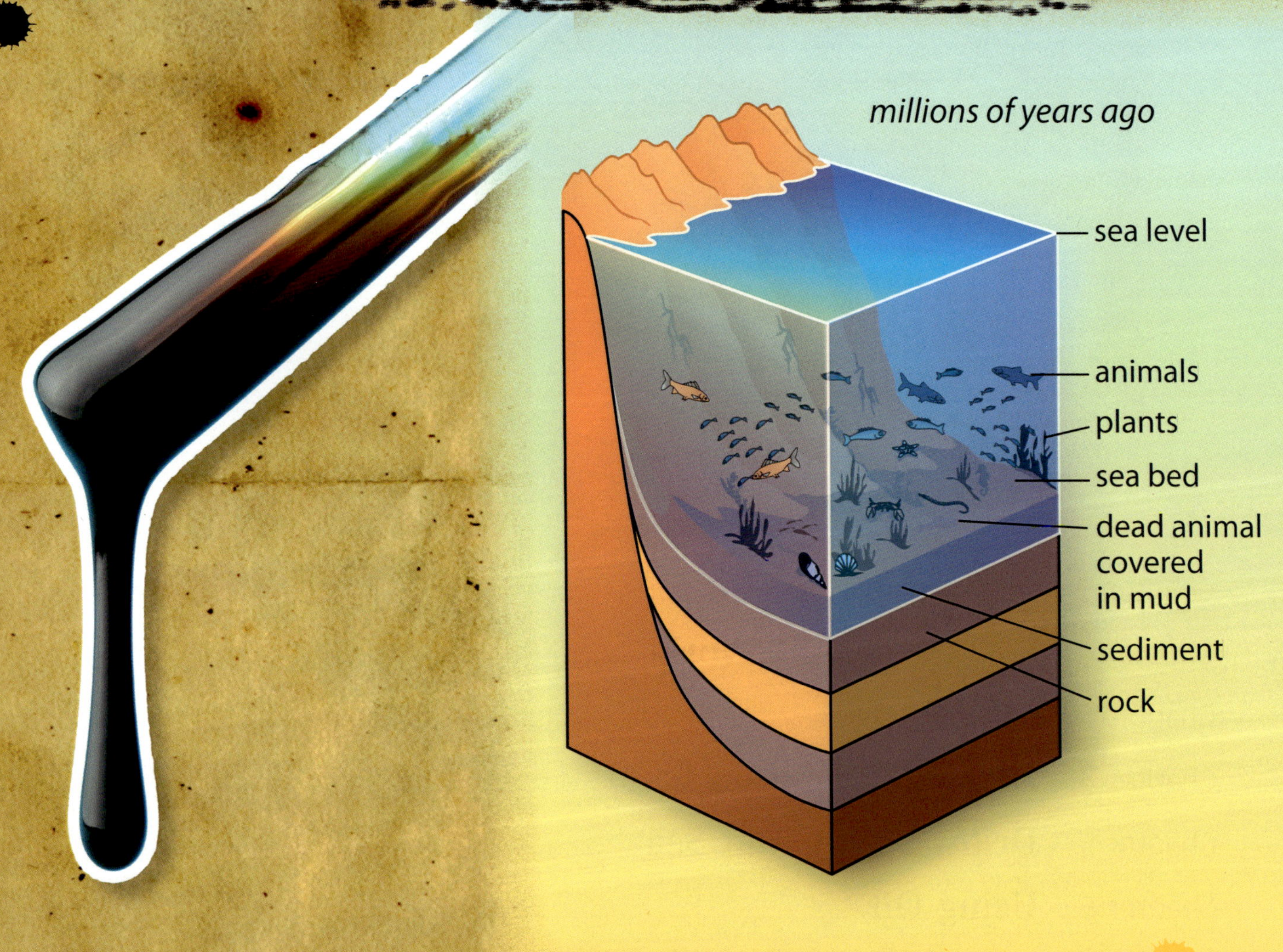

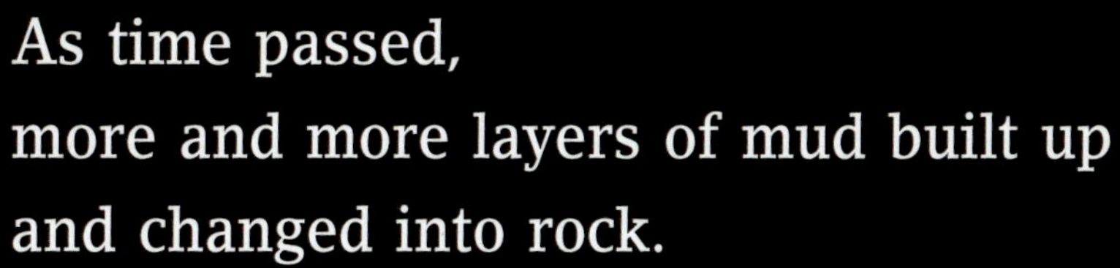

As time passed,
more and more layers of mud built up
and changed into rock.

The animal and plant remains became **fossils**.
After millions of years,
the fossils turned into oil and gas.
This is why oil is called a fossil fuel.

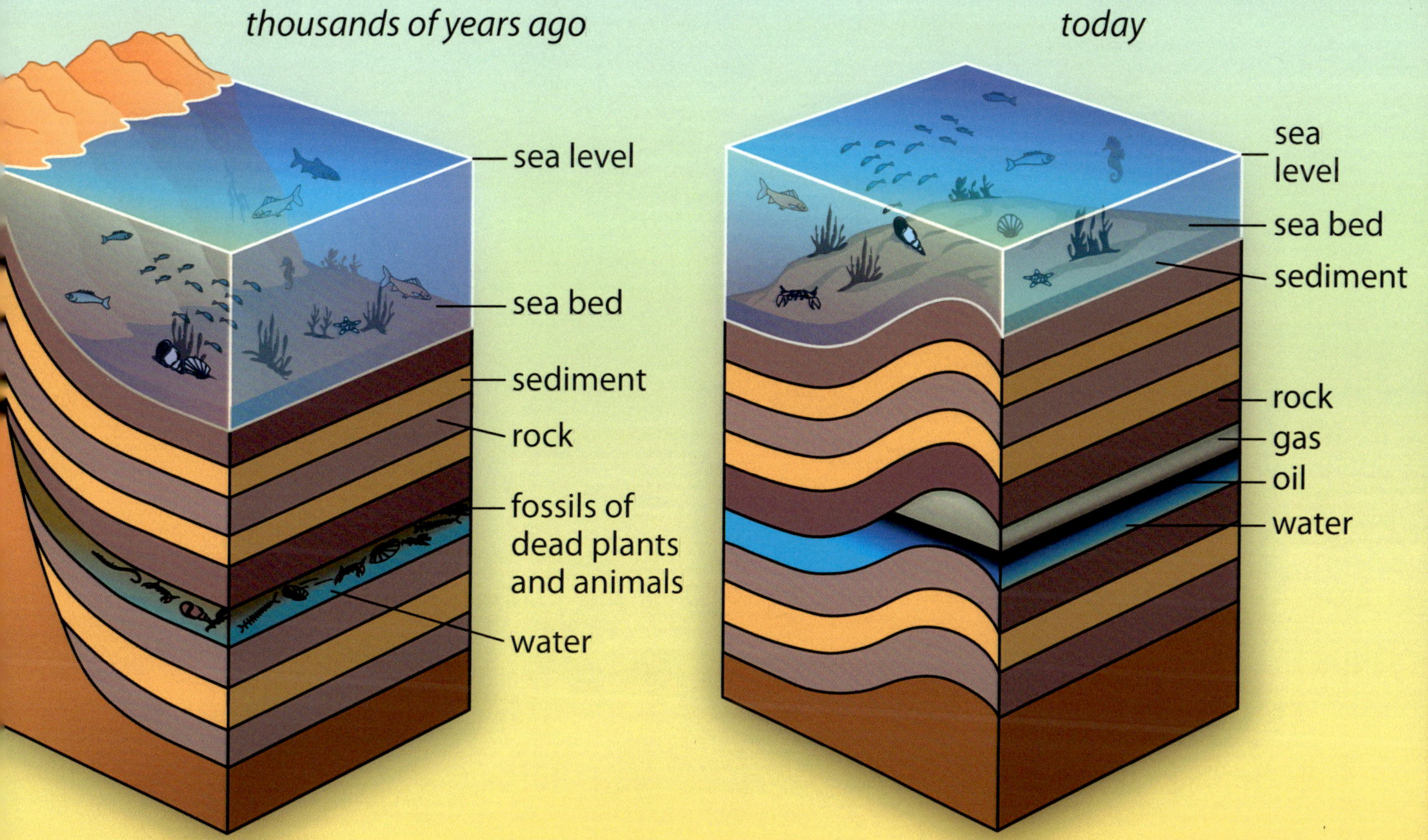

thousands of years ago

sea level
sea bed
sediment
rock
fossils of dead plants and animals
water

today

sea level
sea bed
sediment
rock
gas
oil
water

Dead plants and animals drop to the sea bed and are covered with a layer of mud.

CHAPTER 2

Oil Around the World

Many places on Earth
that were once covered by the sea
are now dry land.
This happened because of the movement of Earth's crust
over millions of years.
So there are **oil reservoirs** beneath the ground
in some countries,
as well as beneath the ocean floor.

an oil field in the Taklimakan Desert, China

Sometimes movement of Earth's crust brings oil reservoirs to the surface. In some parts of the world, the oil appears on the ground like a lake.

In the Caribbean countries of Trinidad and Tobago, lakes of oil have appeared on the surface of the land.

The first oil fields were discovered in Canada
more than 100 years ago.
Now most of the world's oil fields have been found.
It is becoming harder and harder
to find new oil reservoirs.
Oil companies are looking for new oil fields
in remote places like Alaska and the African desert.

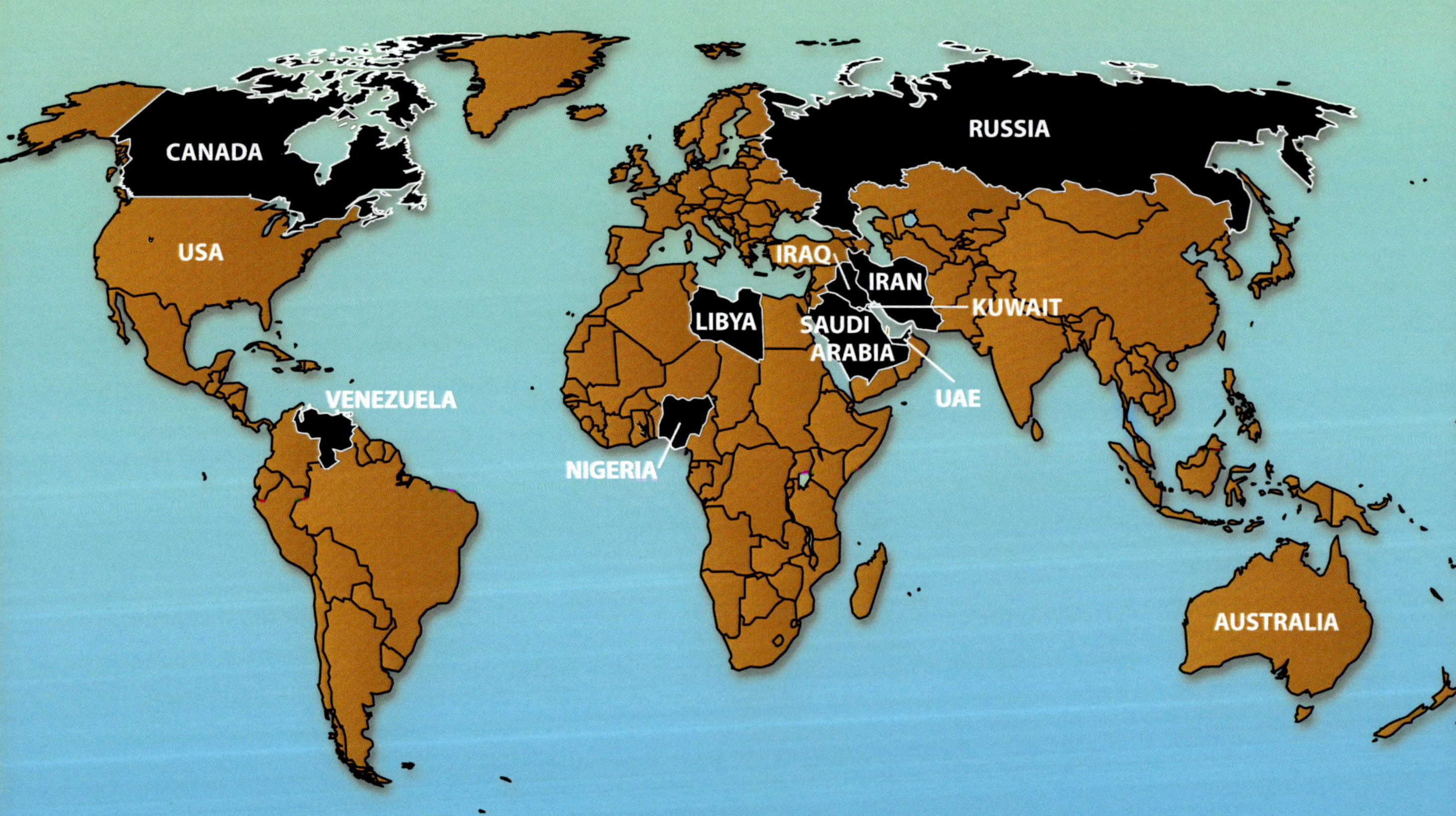

The countries with the largest amounts of oil are Saudi Arabia, Canada, Iran, Iraq and Kuwait.

Since the 1950s,
oil has become the world's most important source of energy.
Some countries have gone to war over oil
because it is so valuable.

In 1990, Iraq invaded Kuwait to take over its oil fields. Iraq set many of Kuwait's oil fields on fire when it lost the war.

CHAPTER 3

Finding Oil

Scientists try to work out
where oil might be found by studying rocks.
There are special signs that help them work out
whether or not an area contains oil.

Different types of rocks have different levels
of **magnetic** pull.
Rocks that contain oil have a weak magnetic pull.
Scientists use special equipment
to test the magnetic pull of rocks.

A magnetometer is used to measure the strength of magnetic fields.

A gravimeter is used to measure small changes in the pull of gravity.

Another sign is the **gravity** around rocks. Scientists measure the pull of gravity on a rock. Rocks with oil have less gravity around them.

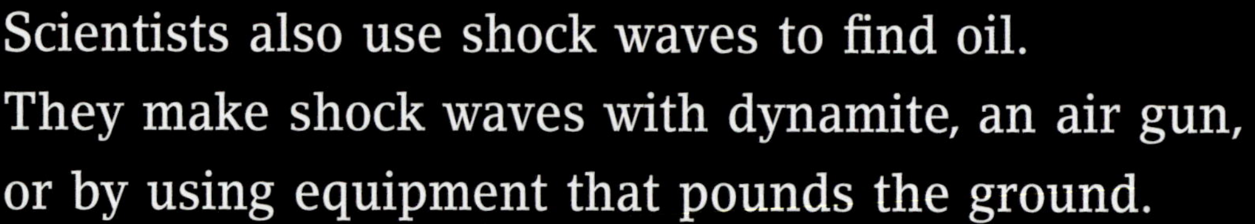

Scientists also use shock waves to find oil. They make shock waves with dynamite, an air gun, or by using equipment that pounds the ground.

The shock waves travel beneath Earth's surface and are reflected back by various rock layers. The reflected sound waves are used to work out if there are any signs of oil.

If these tests suggest that an area might contain oil, the scientists drill down to find it. The scientists test the oil and make a decision about whether it is worth building an **oil rig** there.

A seismogram is a record of sound waves. The results can indicate that oil is underground.

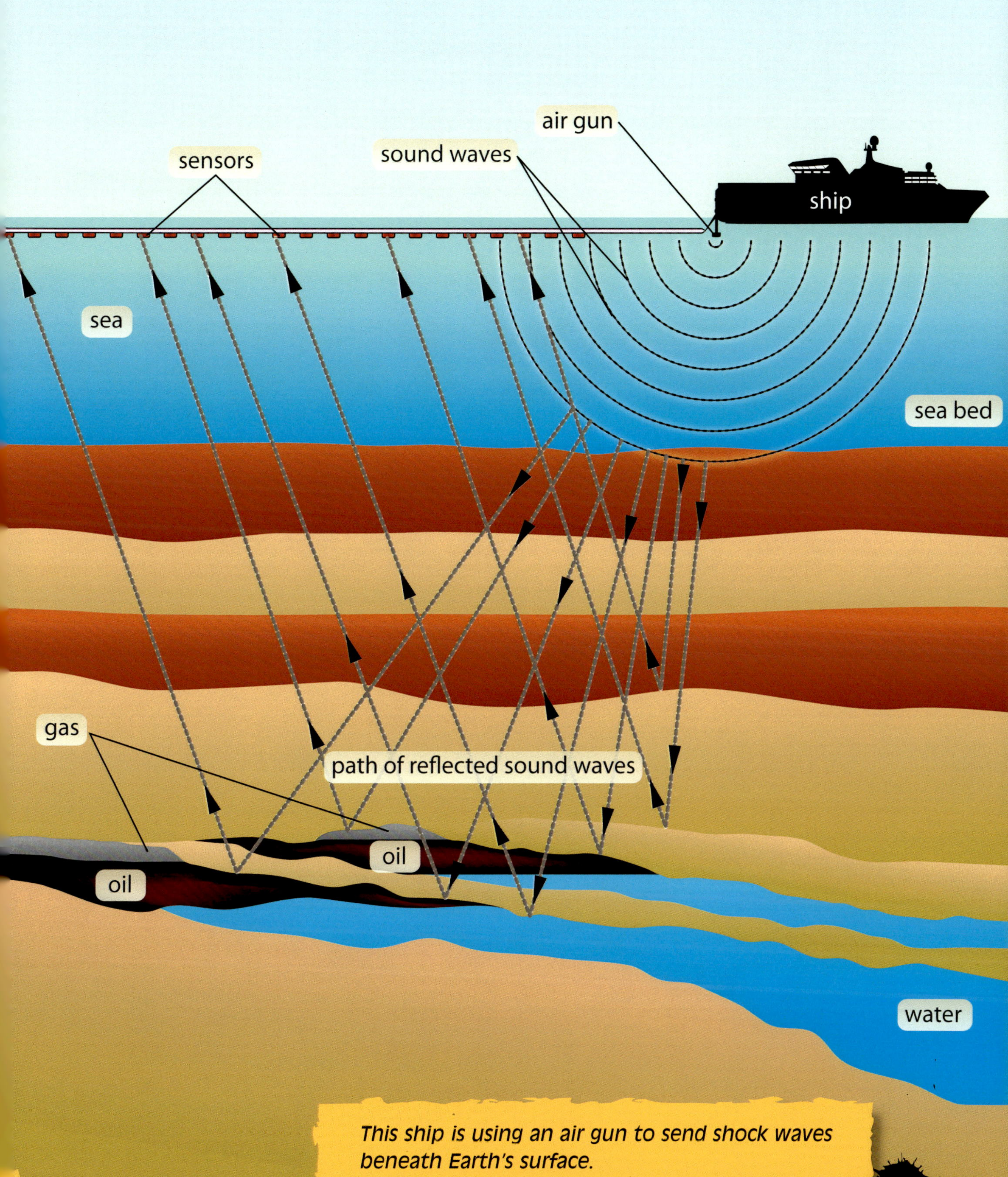

This ship is using an air gun to send shock waves beneath Earth's surface.

CHAPTER 4

Drilling for Oil on Land

After a drilling site on land is chosen,
the land is cleared and made flat.
Roads have to be built to get workers
and equipment to the site,
and water and power must be organised.

Bulldozers clear the trees and flatten the land to make way for a drilling site.

an oil rig

Homes for the workers and other buildings need to be built, and the oil rig is set up.

The oil rig is the machine used to get the oil out of the ground.
It is made up of many parts.

The engine makes the drill string turn fast
down into the ground.

The drill bit cuts underground rock.
As it cuts deeper into the ground,
pieces of steel pipe are added to the drill string.
The drill string can be up to 7600 metres long.
Sometimes the oil rig workers need to change the drill bit
when it is no longer sharp.

Once oil is reached,
it is pumped back up through the pipes of the drill string
into a pipeline or into tanks.

These pipes carry oil to the storage tank.

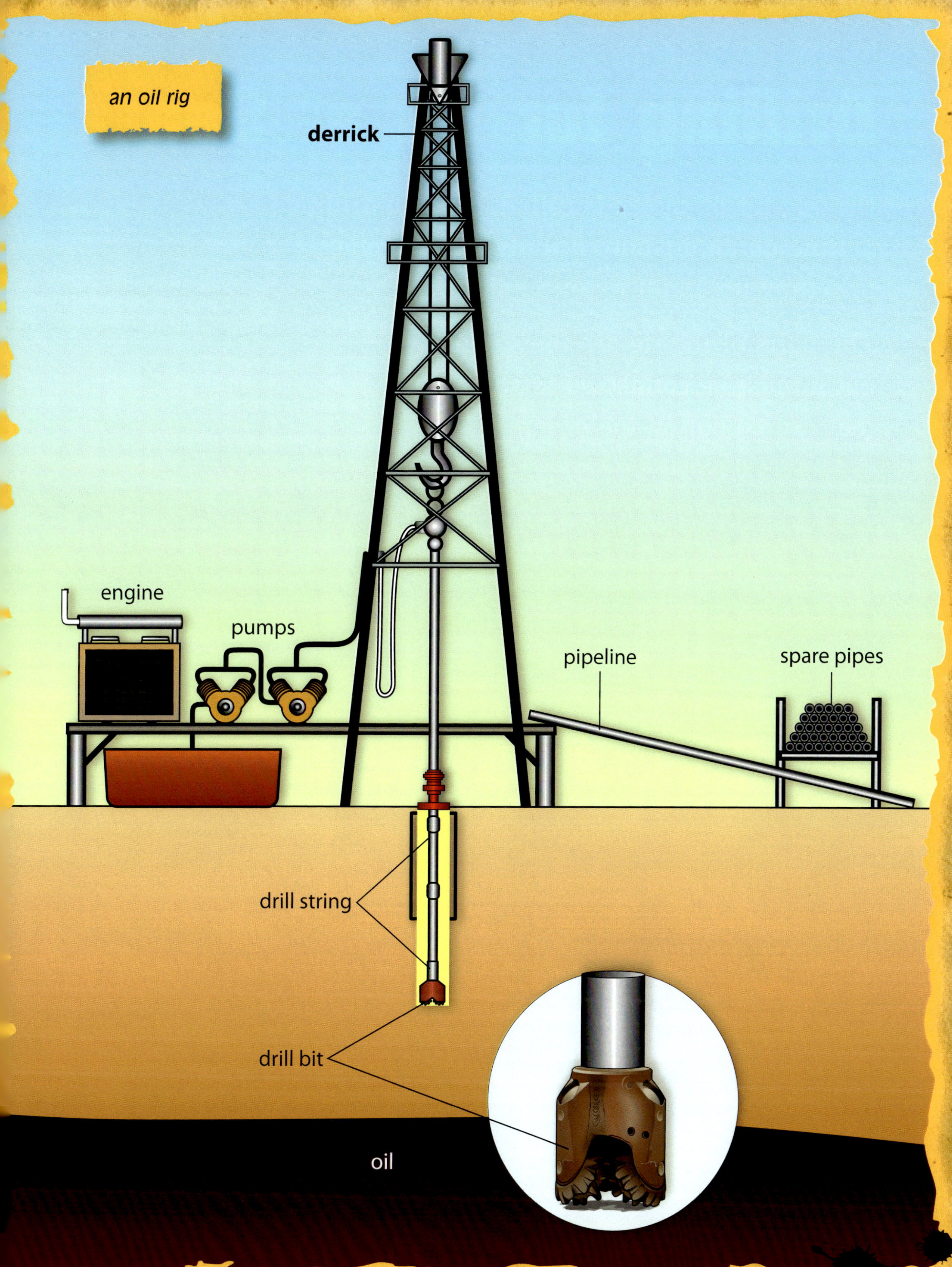
an oil rig
derrick
engine
pumps
pipeline
spare pipes
drill string
drill bit
oil

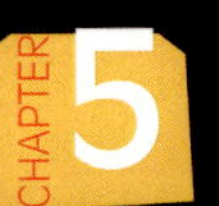

CHAPTER 5

Drilling for Oil at Sea

Drilling for oil at sea is called offshore drilling. Setting up an offshore drilling rig is similar to setting up an oil rig on land.

The parts of the rig are the same, but it must be connected to something that can go out on the ocean.

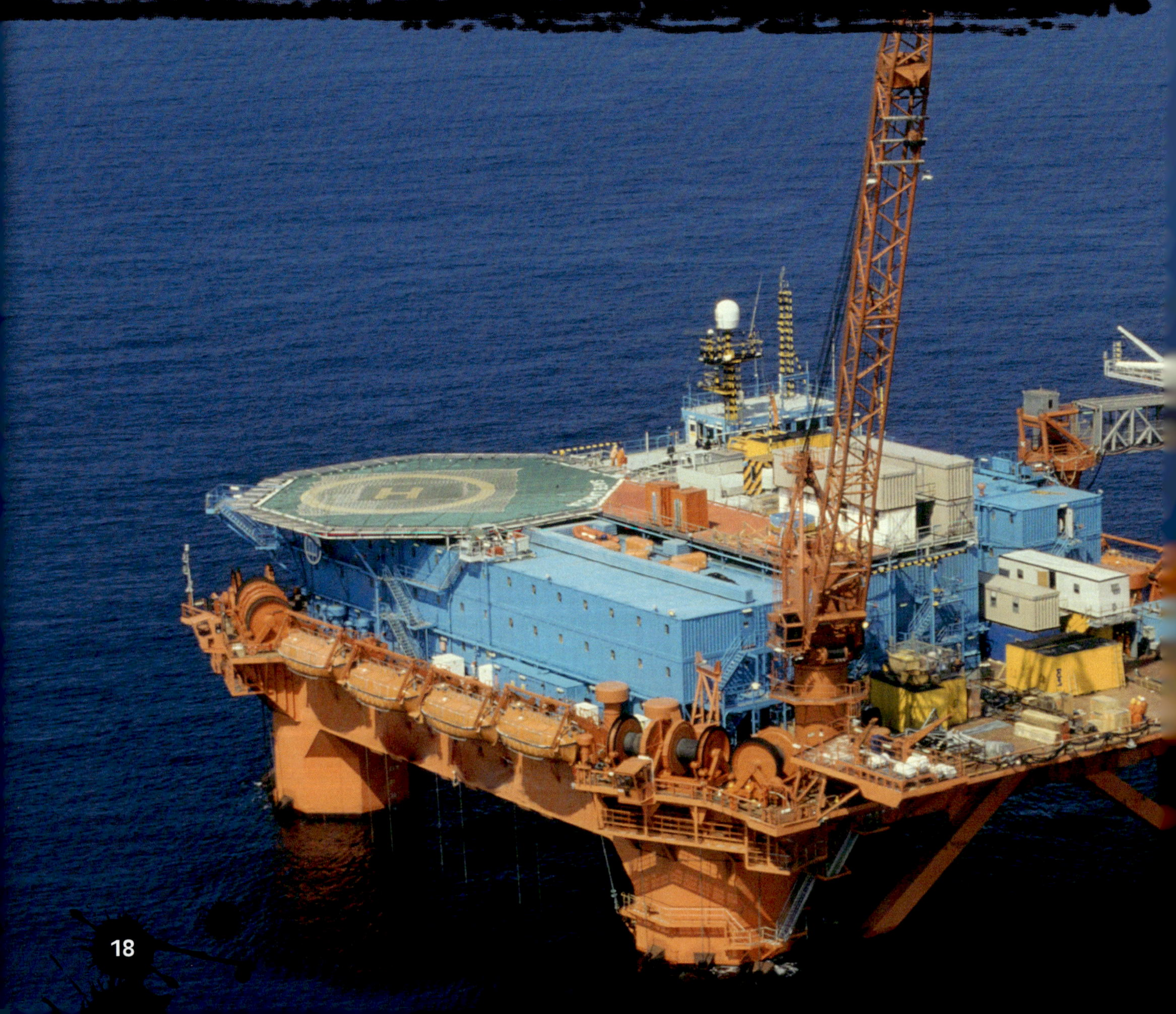

There are different ways of drilling for oil
from beneath the ocean floor.
It depends on how deep the water is,
how big the oil field is,
and how far the site is from the shore.

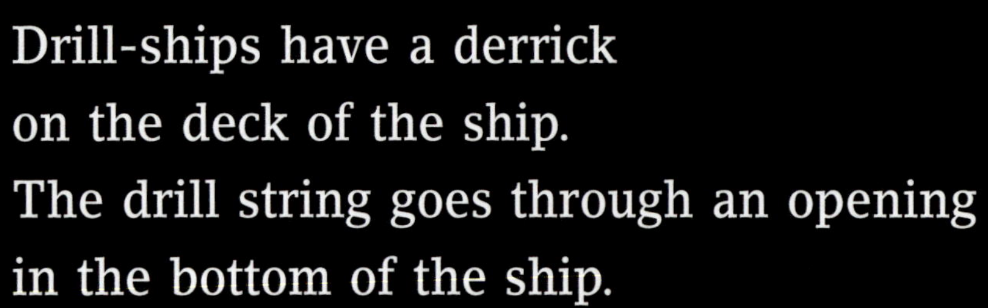

Drill-ships have a derrick
on the deck of the ship.
The drill string goes through an opening
in the bottom of the ship.

Computers on the ship make sure that the ship stays
in exactly the same position over the drilling site.

Drill-ships can drill as deep as 2.4 kilometres below the sea.

Some rigs are built on a floating platform
with steel legs that can extend
down to the ocean floor.
These are called jack-up rigs.
When the legs are lifted up,
a boat can move the rig to a different site.

Production platforms are fixed in the water.
They are very expensive to build.

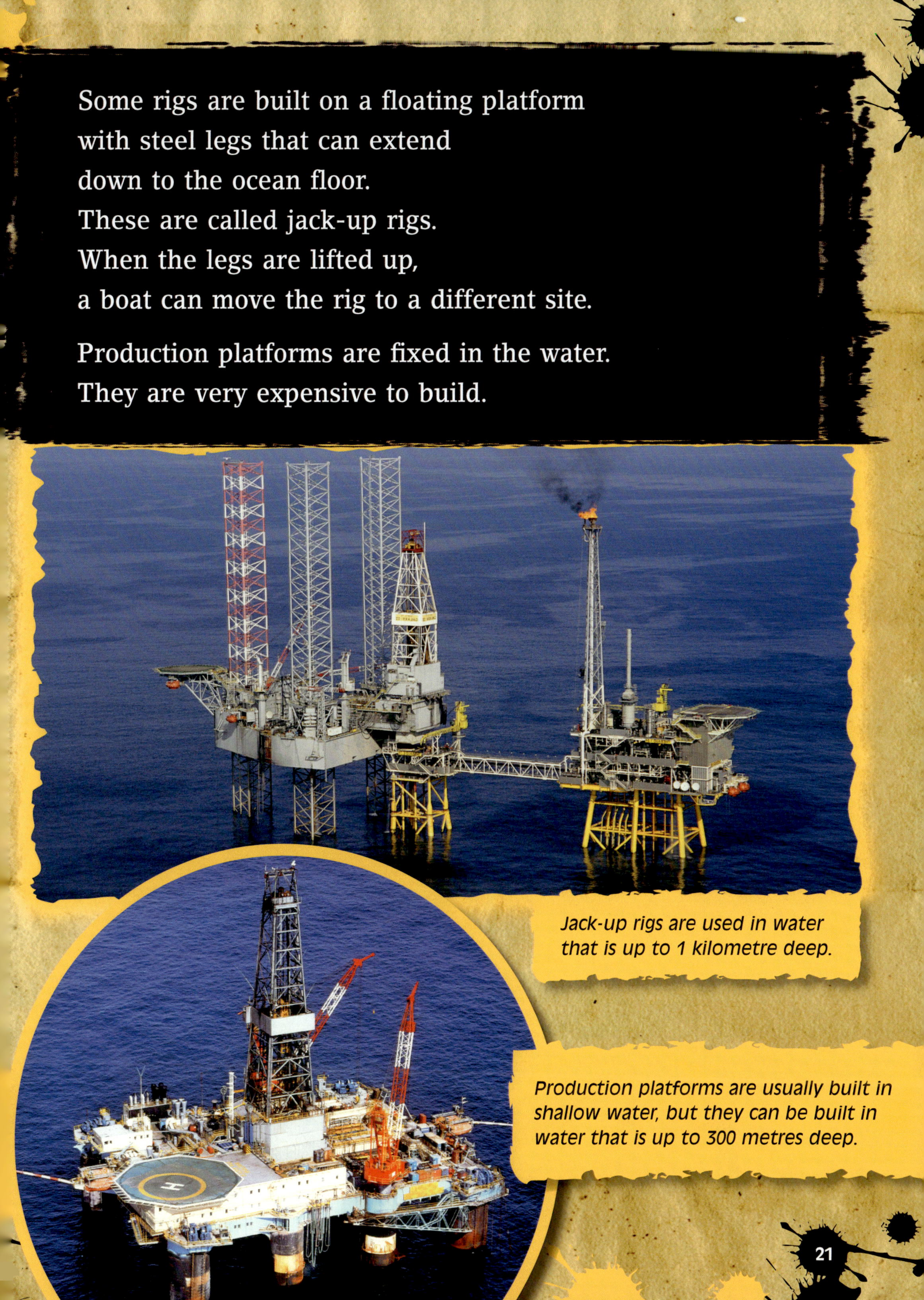

Jack-up rigs are used in water that is up to 1 kilometre deep.

Production platforms are usually built in shallow water, but they can be built in water that is up to 300 metres deep.

CHAPTER 6

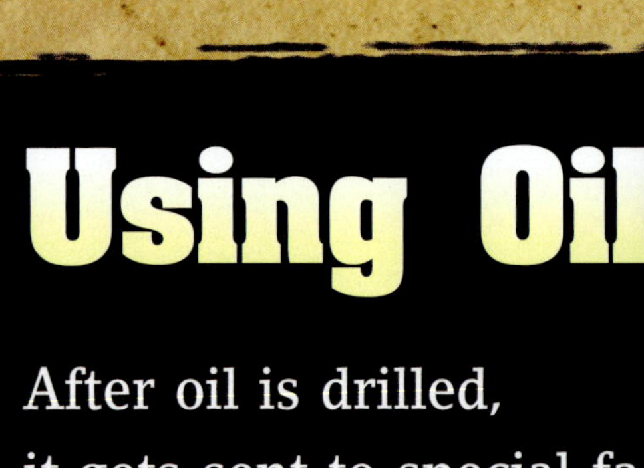

Using Oil

After oil is drilled,
it gets sent to special factories
where it is processed
and turned into useful products.

An oil refinery is where oil is processed.

Many different materials are made from oil,
including fuel for cars, jets,
buses and trucks,
and raw materials for plastic.

polystyrene cups

This plane is leaving trails of exhaust as its engine burns oil.

Lots of research and hard work goes into finding oil, drilling it from the ground and turning it into products that we use every day.

But Earth's oil supply will not last forever. It is very important to find an alternative before Earth's oil runs out.

Glossary

derrick the part of an oil rig that raises and lowers the drill

fossils any trace of plants or animals from another time

gravity a force that stops everything on Earth from drifting off into space

magnetic the force of attraction between all particles

oil reservoirs pockets of oil trapped deep beneath layers of rock

oil rig the machine that gets the oil out of the ground

remains the body or skeleton of a dead person, animal or plant

Index